Waste Plastics

Topical Conference at the 2022 AIChE Annual Meeting

Phoenix, Arizona, USA
13-18 November 2022

ISBN: 978-1-7138-7909-1

Printed from e-media with permission by:

Curran Associates, Inc.
57 Morehouse Lane
Red Hook, NY 12571

Printed with permission by Curran Associates, Inc. (2023)

For permission requests, please contact AIChE
at the address below.

AIChE
120 Wall Street, FL 23
New York, NY 10005-4020

Phone: (800) 242-4363
Fax: (203) 775-5177

www.aiche.org

Additional copies of this publication are available from:

Curran Associates, Inc.
57 Morehouse Lane
Red Hook, NY 12571 USA
Phone: 845-758-0400
Fax: 845-758-2633
Email: curran@proceedings.com
Web: www.proceedings.com

TABLE OF CONTENTS

ANALYSIS AND ASSESSMENT FOR SOLVING THE PLASTIC WASTE CRISIS

CONVERSION OF WASTE PLASTIC INTO LIQUID FUELS

PLENARY SESSION: WASTE PLASTICS (INVITED TALKS)

RECYCLING AND UPCYCLING OF PLASTIC WASTE

576a Leveraging Polymer Characterization to Gain Key Insights into Polyolefin Deconstruction Processes

Zachary Hinton, Thomas H. Epps III, LaShanda Korley

New technologies to mitigate the accumulation of plastics waste provide an opportunity to better utilize valued carbon resources. The diversity of valorization, recycling, and upcycling strategies has spearheaded research that optimizes variables, such as (bio)catalyst type or features, product selectivity, and process design. Unfortunately, critical concepts (e.g., transport phenomena, thermodynamics) and experimental characterization (e.g., rheology, calorimetry) for macromolecules often are not integrated into end-of-life plastics management. In this work, we present two case studies that connect polymer physics and methodology to the deconstruction products obtained from plastics waste. In the first study, we demonstrate the effect of polyethylene architecture on conversion via hydrotreatment. We show that the fundamental kinetic and transport parameters associated with deconstruction reactions can be explained using rheological measurements of the polymer melt. Although at the monomeric level all polyethylenes are functionally equivalent, we show that fundamental polymer features (e.g., branches, entanglements) are responsible for important differences in catalytic performance in deconstruction processes. In the second case study, we highlight the importance of large molecular weight (polymeric) products generated by the catalytic deconstruction of polyolefins and polystyrene in understanding the overall valorization process. Using detailed mathematical analysis of gel permeation chromatography data, we quantify product species that are often hidden by unreacted polymer, providing insight into kinetic and transport mechanisms associated with polymer deconstruction. Overall, this work presents a framework for important assessments of waste plastic conversion strategies using first principles to drive meaningful innovations in plastics waste valorization.

576b Microplastics in Foods and Beverages

Robert Peters, Larry Erickson

There is a growing interest in the plastic content in foods and beverages because of small particulates in air, water, and soil that contain synthetic chemicals. Some of these particulates, which often contain chemicals and microorganisms of concern, find their way into foods and beverages. Microplastics (plastics with size less than 5 mm in diameter) impact product quality and consumer behavior. The importance of addressing issues associated with microplastics in food and beverage manufacturing and processing is increasing because of greater emphasis on plastics in the environment and concerns about impacts on human health. Impacts of microplastics on aquatic ecosystems will also be addressed. The results of a review of the literature will be presented.

576d Integrated Life Cycle and Techno-Economic Analysis Framework for Hydrothermal Liquefaction of Municipal Waste

Seshasayee Mahadevan Subramanya, Phillip E. Savage, Rui Shi

We developed an integrated life cycle assessment (LCA) and techno-economic analysis (TEA) framework to elucidate the sustainability implications of hydrothermal liquefaction (HTL) as a valorization technology for municipal solid waste. Component additivity models were integrated with LCA and TEA modeling to provide mechanistic links among unit operations and to perform dynamic cost and life cycle environmental impact assessments of processing various feedstock compositions at different HTL temperatures. Displacement credits were applied for co-produced biochar, terephthalic acid, and bisphenol-A. Uncertainties were characterized through Monte Carlo simulation, and global sensitivity analyses were conducted using both Spearman correlation and Morris method to identify the key drivers for economics and environmental impacts. Our results showed that the simulated municipal waste in the New York metropolitan area can be processed via HTL (at 425 °C) with global warming potential of – 0.60 ± 0.35 kg CO_2 eq. per kg feedstock processed using displacement allocation and produce gasoline – diesel blend with minimum fuel selling price of 1.48 ± 0.51 \$/gallon. We also compared the HTL system with conventional waste processing scenarios to identify the suitability of feedstocks for HTL. The results indicate HTL having a potential GHG saving of 0.33 kg CO_2 per kg MSW processed when compared to the current practice.

576e Progress Towards to a Circular Plastic Economy: Intrinsic Kinetics of Polypropylene Pyrolysis Via Pulse-Heated Analysis of Solid Reactions (PHASR)

Nathan Sidhu, Isaac Mastalski, Paul Dauenhauer

Polymeric materials have become ubiquitous in modern life, as evidence by the rapid and continual growth in global plastic production. Consequently, billions of tons of plastic waste have been produced which are insufficiently managed under the current linear plastic economy wherein the majority of plastic waste has been discarded in landfills or the environment. A material of principal importance is polypropylene, one of the most widely produced thermoplastic materials, accounting for approximately 20% of global polymer production. Under a circular plastic economy, end-of-life plastics are instead recycled in a closed loop, fully regenerating the original polymers. To realize a circular plastic economy, new recycling techniques must be developed. Pyrolysis, the thermal degradation of a material in an inert atmosphere, has high potential as a driving technology for a circular plastic economy. Quantification of the intrinsic kinetics of plastic pyrolysis is an ongoing a challenge, owing to the complexity of the pyrolysis mechanisms and limitations of standard analytical methods.

In this work, the novel 'pulse-heated analysis of solid reactions' (PHASR) reactor system has been utilized to study polypropylene pyrolysis. The PHASR reactor, originally developed for cellulose pyrolysis and redesigned for polyolefin pyrolysis, is uniquely capable of operating absent transport limitations, secondary reactions, or kinetic measurement limitations to measure intrinsic kinetics on the millisecond scale at temperatures up to 700 °C. [1] The abilities of the PHASR reactor to pyrolyze polypropylene under isothermal, kinetically limited conditions are demonstrated in this work and yield vs. reaction time data and intrinsic lumped kinetics are presented. [2] In addition, visual analysis of polypropylene pyrolysis was performed using a Visual PHASR experimental system, which allows for direct observation of pyrolysis reactions via high speed photography. Visualization results reveal reaction phenomena and are compared to experimental results. [2] The data from the PHASR reactor systems provides new insights that will help to enable the design and development of more efficient industrial pyrolysis reactors and ultimately a circular plastic economy.

References:

[1] *ChemSusChem*, **2021**, *14 (19)*, 4214-4227.

[2] *In Preparation*, **2022**.

576f Technoeconomic Analysis of Plastics Recycling Process Via Mechanocatalytic Reactions

Elisavet Anglou, Arvind Ganesan, William Bradley, Sankar Nair, Carsten Sievers, Fani Boukouvala

Conventional plastic recycling pathways such as pyrolysis and solvolysis involve high energy consumption, use of chemicals or solvents and are capital intensive. Such factors currently make recycling economically not viable [1, 2]. Therefore, creating new depolymerization processes to recycle and reuse plastics is crucial in implementing sustainable practices and reducing negative environmental effects. One promising route is the mechanochemical depolymerization of plastic waste wherein the polymer chain is deconstructed by the application of mechanical energy which then activates the chemical reaction. The mechanocatalytic hydrolysis reactions are performed in ball-mill reactors that are known to have the potential to process polymers in the solid state to valuable plastic monomers in a sustainable and economical way[3, 4]. Coupled with the reactor unit, several purification steps are necessary to purify the resulting output stream exiting the ball-mill reactor. To address this challenge, we provide an end-to-end approach for developing a complete process flowsheet for all the unit operations required to transform model polymer feeds (e.g., PET) in powder form to monomers. The unit operations incorporated in the depolymerization process are reactive crystallization, solid-liquid extraction, distillation, and solid filters [5].

In this talk, we will present the model equations, connecting streams and recycle streams incorporated within the integrated model-based flowsheet. Conventional units such as distillation units and solid filters are integrated with custom-made models developed for the ball-mill reactor and reactive crystallization [4-6]. Results of a comprehensive technoeconomic and energy analysis will be presented for each individual unit, and cost-energy trade-offs will be discussed. Finally, sensitivity analysis [7] studies will be performed with respect to the overall achieved yield of the monomer as well as the total amount of raw materials utilized, to identify process degrees-of-freedom that significantly affect cost and energy metrics.

Citations

[1] I. Vollmer *et al.*, 'Beyond Mechanical Recycling: Giving New Life to Plastic Waste,' *Angew Chem Int Ed Engl,* vol. 59, no. 36, pp. 15402-15423, Sep 1 2020, doi: 10.1002/anie.201915651.

[2] R. Meys, F. Frick, S. Westhues, A. Sternberg, J. Klankermayer, and A. Bardow, 'Towards a circular economy for plastic packaging wastes – the environmental potential of chemical recycling,' *Resources, Conservation and Recycling,* vol. 162, 2020, doi: 10.1016/j.resconrec.2020.105010.

[3] V. Štrukil, 'Highly Efficient Solid-State Hydrolysis of Waste Polyethylene Terephthalate by Mechanochemical Milling and Vapor-Assisted Aging,' *ChemSusChem,* vol. 14, no. 1, pp. 330-338, 2021.

[4] J. Lu, S. Borjigin, S. Kumagai, T. Kameda, Y. Saito, and T. Yoshioka, 'Practical dechlorination of polyvinyl chloride wastes in NaOH/ethylene glycol using an up-scale ball mill reactor and validation by discrete element method simulations,' *Waste Management,* vol. 99, pp. 31-41, 2019/11/01/ 2019, doi: https://doi.org/10.1016/j.wasman.2019.08.034.

[5] M. A. McDonald *et al.*, 'Reactive crystallization: a review,' *Reaction Chemistry andamp; Engineering,* vol. 6, no. 3, pp. 364-400, 2021.

[6] N. Metta, M. Ierapetritou, and R. Ramachandran, 'A multiscale DEM-PBM approach for a continuous comilling process using a mechanistically developed breakage kernel,' *Chemical Engineering Science,* vol. 178, pp. 211-221, 2018/03/16/ 2018, doi: https://doi.org/10.1016/j.ces.2017.12.016.

[7] A. Saltelli, M. Ratto, S. Tarantola, and F. Campolongo, 'Sensitivity analysis for chemical models,' *Chemical reviews,* vol. 105, no. 7, pp. 2811-2828, 2005.

576g A Platform for Rapid Portable Analysis of Environmental Micro/Nano-Plastics

Eric Johnston, Victor Ugaz

Plastic waste pollution poses significant concerns because polymeric materials rapidly degrade into micron to nanometer-sized particulate fragments. These tiny micro/nano plastic debris become easily dispersed throughout the environment where they enter the food supply and pose potentially significant health risks. But current analysis methods cannot differentiate and quantify polymeric nanoparticles (i.e., nanoplastics) in the ocean and freshwaters from other types of nanoparticles. Here we address this need by applying a microfluidic approach that enables nanoplastic particles to be characterized by size and concentration. Two laminar streams, one containing a tracer dye solution and the other containing an aqueous polymeric nanoparticle suspension, are injected in parallel through a microchannel. The dye and nanoparticles laterally diffuse across the interface between the two steams, where the resulting interfacial complexation produces a unique fluorescent signature distinct from the diffusion profile expected from either the dye or nanoparticles alone. A ratiometric analysis approach correlates the difference between the maximum and minimum of the first-order spatial derivative of fluorescent intensity to the nanoparticle surface area available for complexation with the tracer dye. We demonstrate the operation of our system using polystyrene nanoparticles in the 20 – 100 nm size range to establish the fluorescent signal's dependence on size and concentration, and quantify sensitivity and detection limits. The choice of tracer dye can be tuned to produce fluorescent signatures that depend on nanoparticle composition, laying the groundwork for rapid field-based analysis of environmental micro/nano plastics.

530a Effect of Acidic Pyrolyzed Hydrochar on Hydrothermal Liquefaction of Waste PVC

Vahab Ghalandari, Toufiq Reza

The United States generates the most waste plastic waste per capita in the world, with the average American generating 130 kg each year (42 million metric tons). Waste Polyvinyl chloride (PVC) is one of the main portions of waste plastic, which is a halogenated plastic listed as number 3 in the recycle class. Halogens possess a significant risk to waste-to-energy as well as any other thermochemical conversion- resulting in landfilling the entire waste PVC. Among other dehalogenation pathways, hydrothermal pathways have been recently proposed. In subcritical conditions, water reacts with PVC and leaches chloride from the solid. Hydrothermal reactions are often catalyzed by acids. In this study, the effect of hydrothermal liquefaction (HTL) of waste PVC is investigated in presence of acidic pyrolyzed hydrochar. Pyrolyzed hydrochar was prepared by acidic hydrotreatment carbonization of pineapple waste and resulting hydrochar was pyrolyzed at 400 °C in the presence of CO_2. Pyrolyzed hydrochars are acidic, stable, porous, and containing acidic functional groups. Pyrolyzed hydrochar was co-fed with PVC during HTL to enhance conversion, dehalogenation, and crude quality. HTL experiments were performed using a Hast-C Parr batch reactor at 300-350 °C, 0.25-4 h of reaction times, and 0-20% char to PVC ratio. The results showed that pyrolyzed hydrochars enhance HTL conversion with a maximum value of 24.7 wt% at 5 wt% char content at 350 °C and 0.5 h. Furthermore, crudes included far less chloride and significantly more carbon and hydrogen, resulting in a significantly greater heating value. Finally, the results demonstrated that moderate HTL conditions can efficiently remove chlorine from feedstock and shift it into the aqueous phase, demonstrating the potential of converting PVC to clean solid fuel using HTL.

530b Catalytic Microwave-Assisted Pyrolysis of Plastic Waste for Fuels and Carbon Material

Leilei Dai Sr., Roger Ruan Sr., Paul Chen, Yanling Cheng

Plastic materials are extremely popular all over the world and used in the different walks of life due to the inherent properties of being strong, lightweight, and easily shaped. However, the vast majority of waste plastics ever produced enters into landfills or our ecosystems, creating a plastic waste crisis. Waste plastic conversion for the production of green fuels, chemicals, and hydrogen is able to reduce the consumption of fossil energy, thereby alleviating the climatic effects and achieving the 1.5 °C Paris climate goal. Currently, the most common solution to recycle the plastic waste is mechanical method that is limited by the corresponding challenges. The recycled plastics after melting and remolding generally have poorer properties than those of the virgin plastics, enabling the recycled plastics to be only applicable to lower quality products. This downcycling process makes the waste plastic recovery not attractive in the industry. In this regard, we proposed a novel catalytic microwave-assisted pyrolysis (CMAP) technology that is designed to achieve the conversion of waste plastics to high quality naphtha, that are injected to new plastic manufacturing, with the hope of creating a circular economy and minimizing greenhouse emission in mind. The byproduct, non-condensable gas, ever produced from the pyrolysis-reforming process will be recovered for carbon material and hydrogen via chemical vapor deposition technique.

Recently, we have screened different catalysts to improve the yield and quality of the liquid hydrocarbons from plastic waste pyrolysis, with the aim of maximizing naphtha fractions for new plastic production and extending the catalyst lifetime as much as possible. First, we proposed a novel approach of combining two catalytic reforming zones, where the first catalytic reforming zone was intended to improve the cracking of polyolefins into short chain olefins, and the second, lower-temperature catalytic reforming zone was intended for the hydrogenation process to convert these olefins to C5-C12 alkenes. The tests were very successful and very promising results (60-75% C5-C12 alkanes, 3-5% C5-C12 olefins, 5-15% mono-aromatics) were obtained as hypothesized. Unfortunately, the catalysts for plastic cracking are prone to be deactivated easily. The relationship between the catalyst structure and catalyst lifetime was studied comprehensively by fine-tuning the acidity and pore structure of zeolites. It was confirmed that proper acid density and larger pore size significantly improved the catalyst lifetime. Considering that the pore size of conventional ZSM-5 catalyst is too small for plastic pyrolysis intermediates entering into the pore system, which will limit the diffusion of intermediates and block the pore opening and result in the fast deactivation of catalyst, the hierarchically macro-meso-microporous high Si/Al ratio ZSM-5 zeolite was developed and tested. It was confirmed the catalyst lifetime of hollow ZSM-5 is five times longer than the conventional ZSM-5 due to the better diffusion channels and improved accessibility of acid sites inside catalysts. After producing fuels or chemicals, 30~40 wt.% non-condensable gases (mainly composed of hydrogen and C1-C4 hydrocarbons) will be left, to be further treated. So, we synthesized a metal oxide coating material to catalytically decarburize the non-condensable gases for carbon material and hydrogen production. When the non-condensable gas products pass through a high temperature reactor packed with an effective Ni-Fe/Al_2O_3 catalyst, high yield of hydrogen (88% hydrogen selectivity) can be achieved, with carbon products being produced. In summary, our studies indicated that the CMAP is a low-cost and highly efficient technology convert waste plastics to naphtha and hydrogen.

530c Hydrothermal Liquefaction of Waste Plastics and Lignin

Sampath Gunukula, Sampath Karunarathne, Clayton Wheeler, Somtochukwu Anonyuo

Lignin is the second most abundant biopolymer in nature and is currently underutilized. Plastic waste has increased considerably after the onset of coronavirus pandemic. Therefore, it is desirable to find a sustainable solution for increasing the uses of underutilized lignin and plastic beyond incineration for energy production. In this work, we study the synergetic effects of depolymerizing lignin and plastic (polyolefins) mixture to transportation fuel intermediates using hydrothermal liquefaction. We hypothesize that supercritical water and aromatic mixture produced from lignin depolymerization can cause a high dissolution of molten polyolefin phase, which eventually can promote polymer homolysis and unimolecular reactions like beta-scission and surpass coke formation reactions. Further, the hydrogen radicals produced during the beta-scission reactions of a polyolefin can react with reactive intermediates of lignin depolymerization under supercritical conditions and produce low-molecular-weight aromatic compounds by preventing the coke forming condensation reactions. We further show the effect of lignin chemical structure and branching structure of polyolefins on the quality of produced fuel intermediates under supercritical water conditions. The nuclear magnetic resonance spectroscopy, gel permeation chromatography, and Fourier transform infrared spectroscopy are used to study the chemical characteristics of lignin. Gas chromatography and elemental analysis results will also be presented and will be used to propose a major reaction mechanism of lignin and polyolefin conversion to fuel intermediates under supercritical conditions.

530d Steam Catalytic Conversion of Waste Plastics into Fuels and Chemicals.

Foster Agblevor, Corey Skenandore, Hamza Abdellaoui, Ezra Johnson, Corbin Romney

- Plastics are some of the most useful materials ever evented by man. Plastics are found in consumer products, engineering products, automobiles, airplanes, marine vehicles etc. Since the introduction of synthetic organic polymeric products (resins and fibers) into the consumer market in the early 1950s, the global plastics production has risen from a mere 2 million metric tonnes (Mt) in 1950 to over 380 Mt by 2015. The largest application of plastics is in packaging where the volume has accelerated because of single use containers. Plastics content in municipal solid waste (MSW) has increased from 1% in 1960 to more than 10% by 2005. None of the commonly used plastics are biodegradable and accumulate in landfills or in the natural environment. The most prevalent plastics are: high density polyethylene (HDPE), low density and linear-low density polyethylene (LDPE), polypropylene (PP), polystyrene (PS), polyvinylchloride (PVC), polyethylene terephthalate (PET), polyurethanes (PUR), polyesters, polyamide, acrylic fibers. The pure polymers are usually mixed with additives to produce desirable properties. Despite its usefulness, petroleum derived plastics have become a major pollution problem worldwide glugging gutters, rivers, oceans and landfills. The need to convert waste plastics into useful consumer products and thus reducing environmental pollution cannot be overstated. Current, recycling methods cannot solve the pollution problems. We have investigated the conversion waste plastics (HDPE, PP, PS) into fuels and aromatic compounds. The reactions were conducted under autogeneous pressure reaction conditions in temperature range of 300 to 400 C and reaction times of 15 to 30 min. The catalyst used was a proprietary catalyst formulated in-house. The ratio of water to plastic ratio (W/P) was varied from 0.2 to 10. When W/P ratio was low, the PP reaction products were mostly aromatic compounds, but when the W/P ratio was high the reaction products were mostly linear alkanes. The products from HDPE were mostly alkanes and did not contain much aromatics. The HDPE reactions were strongly affected by the reaction temperature. Analysis of the reaction products showed that the products at high temperatures will qualify as gasoline fraction whereas at lower temperatures the diesel fractions were relatively high. The yields of the liquid products were very high and the gas yield were relatively low. There was hardly any solids production in this process.

530e The Chemistry and Kinetics of Polyvinyl Chloride (PVC) Pyrolysis

Jiayang Wu, Konstantinos Papanikolaou, Feng Cheng, Bennett Addison, Amy Cuthbertson, Gregg T. Beckham, Manos Mavrikakis, George Huber

Pyrolysis is a technology to convert plastic wastes into an oil that can be used for further upgrading into fuels or chemicals. Polyvinyl chloride (PVC) is often found in waste plastics and can cause corrosion problems.[1] During the thermal decomposition of PVC, HCl is unavoidably formed and corrodes the stainless-steel components of the pyrolysis reactor system. One to three wt% of PVC in a waste plastic feed leads to 5,000-10,000ppm of Cl concentration in the crude pyrolysis oil product.[1] The objective of this research is to investigate the kinetics and chemistry of PVC decomposition using combination of experimental and computational modeling to develop a thermal treatment method to remove Cl efficiently.

PVC degradation occurs in two stages, dehydrochlorination and further decomposition of residual PVC. Dehydrochlorination is an autocatalytic reaction that begins at tertiary Cl structural defects. HCl is the main product from this reaction with small amounts of benzene (less than 2.77 wt.% of product formed). The formation rate of benzene has a similar trend as the formation of HCl because HCl can potentially catalyze a homolytic C-C bond cleavage, which gives rise to benzene and an aliphatic fragment. In this work, we studied the decomposition of PVC based on TGA results and used two models in series: a first order model to represent the dehydrochlorination process with the apparent activation energy of 116.7 kJ/mol, and a parallel first order model to represent the decomposition of PVC residue with the main apparent energy of 230.6 kJ/mol. The experimentally determined activation energy was in reasonable agreement with the predicted value for the HCl–catalyzed dehydrochlorination from density functional theory (DFT) calculations. We developed a pretreatment process for PVC and found out that the longer the isothermal treatment held at 320 °C, the more Cl could be removed. The highest Cl removal efficiency reached 99.97%. FTIR and NMR were used to study the structure of treated PVC. The decomposed PVC had around 23% of quaternary carbon, 33% of C=C, 43% of aliphatic carbons, and 1% of primary carbon. The high concentration of the quaternary carbon indicates the treated PVC has a high crosslinked concentration. The most probable structure of the crosslinked center is cyclohexadiene according to our NMR analysis. Our results provide more insights into the PVC degradation chemistry with a proposed lumped PVC degradation mechanism.

1. J. Scheirs, W. K., *Feedstock Recycling and Pyrolysis of Waste Plastics: Converting Waste Plastics into Diesel and Other Fuels*. John Wiley andamp; Sons, Ltd: 2006.

530f Optimization of Waste Plastic to Fuel Oil Plants' Deployment Using Mixed Integer Programming(MIP)

David Muyise

Mixed Integer Programming (MIP) is an approach that involves the optimization of a range of decision variables in order to minimize or maximize a particular objective function. The main objective of this study was to apply the MIP approach to optimize the deployment of waste plastic to fuel oil processing plants in Uganda. The processing plants are meant to reduce plastic pollution by pyrolyzing the waste plastic into a cleaner fuel that can be used to power diesel/ paraffin engines, so as (1) to reduce the negative environmental impacts associated with plastic pollution and also (2) to curb down the energy gap by utilizing the fuel oil. A programming model was established and tested in two case study applications that are, small-scale applications in rural towns and large-scale deployment across major cities in the country. In order to design the supply chain, optimal decisions on the types of waste plastic to be processed, size, location and number of plants, and downstream fuel applications were concurrently made based on the payback period, investor requirements for capital cost and production cost of fuel and electricity. The model comprises qualitative data

gathered from waste plastic pickers at landfills and potential investors, and quantitative data

obtained from primary research. It was found out from the study that a distributed system is suitable for small rural towns whereas a decentralized system is only suitable for big cities. Small towns of Kalagi, Mukono, Ishaka, and Jinja were found to be the ideal locations for the deployment of distributed processing systems whereas Kampala, Mbarara, and Gulu cities were found to be the ideal locations initially utilize the decentralized pyrolysis technology system. We conclude that the model findings will be most important to investors, engineers, plant developers, and municipalities interested in waste plastic to fuel processing in Uganda and elsewhere in developing economies.

530g Liquid-Fed Plastic Pyrolysis Pilot Plant: Effect of Reactor Volume on Product Yields

Daniel Kulas, Ali Zolghadr, David Shonnard

Plastic waste is a global environmental challenge because of poor end-of-life practices. Chemical deconstruction methods such as pyrolysis can convert waste plastic into useful chemical products, providing an incentive for plastic to be recycled instead of ending up in landfills or the environment. In this study, a unique plastic pyrolysis pilot plant was used for fast pyrolysis (1-5 seconds vapor residence time) of HDPE at 600 °C. A dissolution pre-treatment step dissolved the HDPE in recycled pyrolysis wax to create a liquid feed to the pyrolysis reactor, avoiding clogging and bridging of the plastic particles that is common in alternative feeding strategies.

Three different reactor volumes were tested at constant feed rate to determine the effect of residence time on product distribution. Pyrolysis products were cooled through a series of two condenser to separate the products into three groups: a heavy, waxy product (> C15), a lighter liquid product (C6-C15), and a gaseous product (C1-C6). GC/MS analysis was used to characterize the composition of each product. Multiphysics modeling was conducted to estimate the residence time in the reactor and compare the pilot plant results to previously published micropyrolysis research. It was found that residence time had a significant effect on product distribution, with an increase from 1 to 4.5 seconds of residence time causing a 9 wt. % drop in wax (C20-C30) production and a 11 wt. % increase in light oil (C5-C10) production. The trends for product distribution as a function of residence time were found to be consistent between both the pilot plant and micropyrolysis systems, showing that the pilot plant system can be "tuned" to produce the desired pyrolysis product.

422a Plastic Flows and Recycling Processes in the U.S.: Material Recovery Facilities and PET Processes

Raymond Smith

Plastic flows through the economy have increased from approximately 2 Mt/year in the 1950's to 368 Mt in 2019, with prospects for quadrupling further by 2050 [1]. Such large amounts of postconsumer use plastic could be employed in a circular economy, but only if the systems for recycling are in place and used. These system limitations represent barriers to increased recycling of plastics, which this effort examines along with the processes involved and the flows of plastics through the system. The flows of postconsumer use plastic are of interest for application in the U.S. Environmental Protection Agency's report on Advancing Sustainable Materials Management: Facts and Figures [2]. This report includes information on municipal solid waste (MSW) generation, recycling, composting, combustion with energy recovery, and landfilling. Plastics have been identified as components of interest in postconsumer material flows. In this effort, throughputs for material recovery facilities (MRFs) are quantified to model input-output flows and for downstream processing at reclaimers, with the latter preparing the plastics for final reuse. Polyethylene terephthalate (PET) is of particular interest because it is the most widely recycled plastic. PET finds use in plastic bottles, thermoforms for food packaging, carpeting, and textiles, but only the bottles are regularly recycled. As postconsumer PET flows through MRFs and reclaimer recycling processes, the resource use and environmental releases are modeled, along with recycled flowrates and purities which describe the PET recycling system. Ultimately, the recycled PET is converted into a distribution of new products such as fibers, bottles, and films. This recycling system provides an example of a circular economy which, while only partially manifested due to limitations on collection, processing, and use of PET, offers direction on how materials can be maintained in the economy and only downcycled as appropriate [1].

[1] Smith, R.L., Takkellapati, S., and Riegerix, R.C. (2022), "Recycling of Plastics in the United States: Plastics Material Flows and Polyethylene Terephthalate Recycling Processes," *ACS Sustain. Chem. Eng.,* 10, 2084-2096. DOI: 10.1021/acssuschemeng.1c06845

[2] https://www.epa.gov/facts-and-figures-about-materials-waste-and-recycling/advancing-sustainable-materials-management

422b Prospective Issues on the Reuse of Plastics at the End-of-Life Stage in Europe

Baibhaw Kumar, Natalia Cano, Felipe Moura, Gerardo Ruiz-Mercado, Heriberto Cabezas, Csaba Deak

The fate of plastics is a crucial concern for the various facets of environmental well-being. This presentation reviews the prospective issues concerned with plastics at the end of life, primarily in the European region. The bibliometric analysis of the literature reveals the rising interrelation of 'recycling' in academic investigations. So, there is great interest in the issue. The status based on the classification of various forms of plastics with market share in Europe has been analyzed. The growing concern related to non-valorized plastic wastes and their correlation to current import-export data has been presented, depicting severe environmental problems in the future. Plastic reuse avenues could flourish via mechanical, chemical and energy recovery processes. Hence, we further consider the various avenues governing the fate of plastics after they reach the end of life and become waste. Unfortunately, only about 20% of waste plastic enters any recycling stream in the European context. We also study the various barriers existing to the gainful and more sustainable use of plastics at the end of life, and we consider how these could be remedied. But current policies are not enough to support the market requirement and technological readiness of the industries related to plastics recycling. This study will ultimately aim to design supply chain structures and other approaches that can make it possible and economically gainful to recycle more plastics at the end of life in the manner of a circular economy. We are, therefore, looking for solutions to the problem. The study could benefit policymakers, researchers, and entrepreneurs to understand various facets of plastic recycling in the European region.

422c Finding Strategic Solutions to Waste Plastics, Recycling and Beneficial Reuse Toward Circular Economy

Tapas K. Das, Robert Peters

Over the last 70 years, plastic usage has dramatically increased, with applications in every aspects of modern life. Plastics have transformed our society in innumerable ways and are an integral part of nearly every commercial, business and household. Some 400 million metric tons of plastics are produced worldwide each year for use in a wide range of applications, from grocery bags and water bottles to biomedical, electronic devices, and military aircraft. As it happens, there's still a great future in plastics – analysts predict that by 2025, production of plastics will quadruple to as much as 1.5 billion metric tons per year. But the ubiquity of plastics and the lack of easy and cheap recycling techniques have resulted in waste plastics and microplastics accumulating in landfills, ocean ecosystems and encroaching on the environment, and leach into our water and food. More than 75% of the plastics produced annually are discarded after single usage. Some of these wastes are incinerated in power plants to produce electricity, which generates greenhouse gases and toxic byproducts. In 2020, about 35 million tons of plastics were produced in the U.S., but only 10% were recycled. The current linear paradigm to produce, consume, and discard after single use involves the need of a complete overhaul. To solve these problems we need to implement a circular economy in which all plastics are recovered and reused or economically recycled after use.

Finding strategic solutions to the plastic waste problem will require innovation in deconstruction and upcycling technologies, the invention of new recyclable-by-design materials, new plastic waste regulations, and public-private partnerships. In this presentation, several case-histories conducted in different countries will be described.

Key References:

Das, T.K. (2005). *Toward Zero Discharge: Innovative Methodology and Technologies for Process Pollution Prevention,* Wiley, Hoboken, NJ.

Rahimi, A. and Garcia, J.M. (2017). "Chemical Recycling of Waste Plastics for New Materials Production", *Nature Reviews Chemistry,* 1, 1-11.

Das, T.K. (2020). *Industrial Environmental Management: Engineering, Science and Policy*, 1e, Wiley, Hoboken, NJ.

Abbott, R. (2021). "Advancing a Circular Economy", Special Issue on "Waste Plastics Recycling" in *Chemical Engineering Progress*, July, 25-41, New York, NY.

422d Global Plastic Pollution Policy Developments for Chemical Engineers

Mary Ellen Ternes

Plastic policy is rapidly development globally with better characterization of harm to human health and the environment from plastic pollution. In the context of the March 2, 2022 United Nations Resolution to End Plastic Pollution, this discussion will review the current status and developing future policy and regulation regarding plastic waste and plastic pollution in the United States and globally, including: a recent history of global plastic waste policy; how plastic waste management fails to prevent plastic pollution, including where United States federal legislation falls short; different approaches other nations, and some states within the United States, have adopted to mitigate plastic pollution; specific aspects of the post-use plastic feedstock that render recycling and reuse problematic, including plastic products that result in plastic pollution that has no route to capture at this time; and global efforts to develop an international agreement to end plastic pollution by 2024. Attendees should gain a better understanding of the potential for future regulation and insight into business risk management approaches for those entities relying significantly on plastic in their business models.

With this presentation you will:

- Learn the current status and potential future policy approaches that may govern plastic products in order to better manage plastic waste and mitigate plastic pollution;
- Understand the fundamental barriers posed by the current heterogenous nature of post-use plastic in reuse and recycling;
- Appreciate the business risk posed by failure to properly manage plastic waste and prevent plastic pollution arising from recent developments in characterizing risk from plastic pollution to human health and the environment.
- Gain insight helpful in developing new approaches that may insulate businesses from the uncertainty of future plastic regulation.

422e Enabling Micro-entrepreneurs to convert plastic waste to Polyfuel using Appropriate Technology

Shelby Browning

Combating the environmental crisis caused by mismanaged plastic waste is a global challenge, especially in developing regions due to a lack of recycling availability and waste management infrastructure. One way communities can combat this challenge is by using the process of slow pyrolysis to convert plastic waste into liquid fuel. This plenary talk will present a novel project started by the US based NGO Empowered Solutions for Environmental Sustainability to enable micro-entrepreneurs to start waste to energy businesses in developing countries. Results from a case study in Uganda will be included.

473a A Techno-Economic and Life Cycle Analysis of Operating a Standalone Material Recover Facility in the United States.

Olumide Olafasakin, Sabrina Bradshaw, Horacio Aguirre-Villegas, Jiaze Ma, George Huber, Victor Zavala, Craig Benson, Mark Mba Wright

The circular economy has become a very popular and relevant field of study since it has been identified as a sustainable solution for depleting resources and environmental degradation [1]. The goal of a circular economy is to reduce resource depletion and encourage sustainable development through recycling and upcycling of materials and energy at the consumer, producer, urban, regional, and national levels [2]. To achieve a circular economy, circular practices that include but are not limited to the recirculation of resources and energy, minimization of resources demand, and the recovery of value from waste must be adopted efficiently [2]. Material Recovery Facilities (MRF) play a pivotal role [3]. The separation of municipal solid waste (MSW) into various streams that have economic value is achieved in the MRFs. This study aims to evaluate the economic feasibility of and greenhouse gases emissions associated with operating a standalone MRF capable of sorting and recovering various recyclables from municipal solid waste.

This study employs a discounted cash flow rate of return (DCFROR) analysis to estimate the net present value (NPV) of recovering recyclables at an MRF. The NPV is a common metric used in evaluating the profitability of investment by finding the current value of future revenues and expenses. This study employs the ISO 14040 and ISO 14044 life cycle assessment framework to evaluate the environmental impacts of processing one metric ton (MT) of MSW in an MRF. The study uses sensitivity and uncertainty analysis to assess the effects of variability in crucial MRF process parameters, MSW composition, and price variability on the NPV and GHG emissions of the MRF. In particular, the study investigates the impacts of various plastic waste components on the MRF profitability and environmental footprint.

Preliminary analysis shows that the cost of building a 120,000 MT/year MRF is about 23 MM USD. The revenue generated from the MRF depends mainly on the composition of the MSW and the prevailing market prices of each recyclable. The DCFROR shows that the NPV across 20 years can range from 3.57 MM USD to 60 MM USD at a 10% internal rate of return. The LCA analysis shows that the emissions can vary from 5.98 to 8.53 kg CO_2 eq emissions depending on the composition of the input MSW. Sensitivity analysis shows that the biggest drivers of the NPV are revenue, operating hours, total fixed capital cost, and MSW tipping fees, while MSW composition, baling wire, and electricity impact the emissions the most. The uncertainty analysis showed a 95% probability of achieving a net positive NPV. Future work will estimate the GHG emissions allocated to each recyclable using various LCA allocation methods.

References:

[1] S. Khan and A. Haleem, "Investigation of circular economy practices in the context of emerging economies: a CoCoSo approach," *Int. J. Sustain. Eng.*, vol. 14, no. 3, pp. 357–367, 2021, doi: 10.1080/19397038.2020.1871442.

[2] V. Prieto-Sandoval, C. Jaca, and M. Ormazabal, "Towards a consensus on the circular economy," *J. Clean. Prod.*, vol. 179, pp. 605–615, 2018, doi: 10.1016/j.jclepro.2017.12.224.

[3] C. Cimpan, A. Maul, M. Jansen, T. Pretz, and H. Wenzel, "Central sorting and recovery of MSW recyclable materials: A review of technological state-of-the-art, cases, practice and implications for materials recycling," *J. Environ. Manage.*, vol. 156, pp. 181–199, Jun. 2015, doi: 10.1016/J.JENVMAN.2015.03.025.

473b Techno-Economic Analysis and Life Cycle Assessment of Modular Microwave-Assisted PET Depolymerization

Yuqing Luo, Esun Selvam, Dionisios Vlachos, Marianthi Ierapetritou

Plastic waste is a pressing global issue due to the massive greenhouse gas emissions associated with its production and the lack of proper end-of-life management.[1] Polyethylene terephthalate (PET) is the most common polymer used for the production of plastics with wide applications in the packaging and textile industry.[2] Currently, only less than 20% of waste PET is recycled in the US, while the rest ends up in landfill or incineration facilities, which causes increasing land and air pollution. Moreover, mechanical recycling inevitably causes product quality degradation, which reduces the value of recycled products.[3] Consequently, chemical recycling and upcycling of PET, such as glycolysis and methanolysis, have become increasingly popular.[4, 5] PET glycolysis utilizes ethylene glycol (EG) to break the polymer chain into bis(2-hydroxyethyl) terephthalate (BHET), an essential intermediate in the industrial PET resin synthesis.[6, 7]

In this work, we evaluated the economic and environmental impacts of a microwave (MW) assisted PET glycolysis process. Unlike most PET glycolysis studies using homogeneous catalysts,[8-10] the proposed process utilizes the heterogeneous ZnO catalyst that is more efficient and easier to separate. First, the flowsheet model of the MW-assisted PET glycolysis process is built using Aspen Plus based on reaction conditions and yields. The traditional BHET production from petroleum-based dimethyl terephthalate (DMT) is also simulated. Next, techno-economic analysis (TEA) is conducted using the discounted cash flow method to calculate the minimum selling prices for both routes. A 'cradle-to-gate' life cycle assessment (LCA) is performed to compare the environmental impacts of MW-assisted PET depolymerization with traditional BHET production from DMT, including the global warming potential, ecotoxicity, and fuel depletion.[11] Our preliminary TEA and LCA demonstrate the economic and environmental benefits of using waste plastics as feedstock.

PET waste is geographically distributed. Thus, building a centralized treatment facility for post-consumer waste PET could suffer from insufficient or disrupted supply and high transportation cost.[12, 13] Hence, we simulated the modular MW-assisted PET depolymerization plant at small scales so that they could be built close to plastic waste source and have faster response to supply uncertainties.[14, 15] Sievers *et al.* proposed a fixed capital cost evaluation method for modular plants with a backbone facility and several expandable production line modules.[14] This method is applied to our modular MW-assisted waste PET depolymerization process for TEA. High PET conversion and BHET selectivity of this process also enable easy separation of the product, leading to reduced production cost even at a much smaller plant capacity than the centralized BHET production facility using traditional petrochemical feedstock.

Reference

1. Zheng, J.; Suh, S., Strategies to reduce the global carbon footprint of plastics. *Nature Climate Change* **2019,** *9* (5), 374-378.
2. Gubbels, E.; Heitz, T.; Yamamoto, M.; Chilekar, V.; Zarbakhsh, S.; Gepraegs, M.; Köpnick, H.; Schmidt, M.; Brügging, W.; Rüter, J.; Kaminsky, W., Polyesters. In *Ullmann's Encyclopedia of Industrial Chemistry*, pp 1-30.
3. Ragaert, K.; Delva, L.; Van Geem, K., Mechanical and chemical recycling of solid plastic

waste. *Waste Management* **2017,** *69*, 24-58.

4. Sarda, P.; Hanan, J. C.; Lawrence, J. G.; Allahkarami, M., Sustainability performance of polyethylene terephthalate, clarifying challenges and opportunities. *Journal of Polymer Science* **2022,** *60* (1), 7-31.
5. Paszun, D.; Spychaj, T., Chemical Recycling of Poly(ethylene terephthalate). *Industrial andamp; Engineering Chemistry Research* **1997,** *36* (4), 1373-1383.
6. Barnard, E.; Rubio Arias, J. J.; Thielemans, W., Chemolytic depolymerisation of PET: a review. *Green Chemistry* **2021,** *23* (11), 3765-3789.
7. Bartolome, L.; Imran, M.; Cho, B. G.; Al-masry, W. A.; Kim, D. H., Recent developments in PET recycling. *Material Recycling - Trends and Perspectives* **2012**, 65 - 84.
8. Troev, K.; Grancharov, G.; Tsevi, R.; Gitsov, I., A novel catalyst for the glycolysis of poly(ethylene terephthalate). *Journal of Applied Polymer Science* **2003,** *90* (4), 1148-1152.
9. Xi, G.; Lu, M.; Sun, C., Study on depolymerization of waste polyethylene terephthalate into monomer of bis(2-hydroxyethyl terephthalate). *Polymer Degradation and Stability* **2005,** *87* (1), 117-120.
10. Wang, Q.; Yao, X.; Tang, S.; Lu, X.; Zhang, X.; Zhang, S., Urea as an efficient and reusable catalyst for the glycolysis of poly(ethylene terephthalate) wastes and the role of hydrogen bond in this process. *Green Chemistry* **2012,** *14* (9), 2559-2566.
11. Guest, G.; Cherubini, F.; Strømman, A. H., Global warming potential of carbon dioxide emissions from biomass stored in the Anthroposphere and used for bioenergy at end of life. *Journal of Industrial Ecology* **2013,** *17* (1), 20-30.
12. Bhosekar, A.; Ierapetritou, M., A framework for supply chain optimization for modular manufacturing with production feasibility analysis. *Computers andamp; Chemical Engineering* **2021,** *145*, 107175.
13. Allman, A.; Zhang, Q., Dynamic location of modular manufacturing facilities with relocation of individual modules. *European Journal of Operational Research* **2020,** *286* (2), 494-507.
14. Sievers, S.; Seifert, T.; Franzen, M.; Schembecker, G.; Bramsiepe, C., Fixed capital investment estimation for modular production plants. *Chemical Engineering Science* **2017,** *158*, 395-410.
15. Lier, S.; Grünewald, M., Net Present Value Analysis of Modular Chemical Production Plants. *Chemical Engineering andamp; Technology* **2011,** *34* (5), 809-816.

473c Combining Chemical and Biocatalysis to Breakdown and Upcycle Polyethylene

Gwendolyn Gregory, Cong Wang, Sunitha Sadula, Raul Lobo, Dionisios Vlachos, Eleftherios Papoutsakis

Combining chemical and biocatalysis to breakdown and upcycle polyethylene

To date, there are few reports of microbial deconstruction or functionalization of the recalcitrant backbone of polyolefins. However, microbes can utilize many of the products from polyolefin deconstruction, including *n*-alkanes. Here we combine chemical catalysis with biocatalysis to valorize polyethylene deconstruction products. High density polyethylene (HDPE) was deconstructed via hydrogenolysis over a carbon-supported catalyst. The resulting *n*-alkane mixture (C_4-C_{35}) was then utilized as a feedstock for environmental microbial consortia collected from a local recycling plant after plastic enrichment. We tested two consortia and found that consortium 37 (C37) could convert ~80% of the alkane mix feedstock after 14 days, while consortium 43 (C43) converted ~40% of the alkane mix. We utilized adaptive laboratory evolution (ALE) to select for enhanced alkane utilization by the two consortia. The consortia were grown on a mixture of *n*-alkane standards as the sole carbon source and passaged three times. The resulting adapted consortia were then tested for utilization of the polyethylene deconstruction products. We found that adapted C37 had minimal improvements as compared to the parent consortia, and in fact had poorer utilization of smaller chain length alkanes. However, adapted C43 had significant improvements in alkane utilization, with a more than 100% increase in utilization of alkanes with C_{20} chain lengths or longer, and a 70% increase in total alkane utilization. We determined using GC-MS that C37 produced hexadecanol, hexadecanoic acid, hexadecyl tetradecanoate and hexadecyl hexadecanoate (wax esters) after growth on hexadecane as a model compound. Wax esters have considerable industrial value, with the longer chain lengths (C_{32}-C_{36}) of highest value. Using environmental consortia, a complex mixture of *n*-alkanes derived from waste polyethylene can be valorized to long-chain alcohols, fatty acids and wax esters. This demonstrates that environmental consortia from plastic-enriched environments can be combined with chemical catalysis to upcycle waste polyethylene.

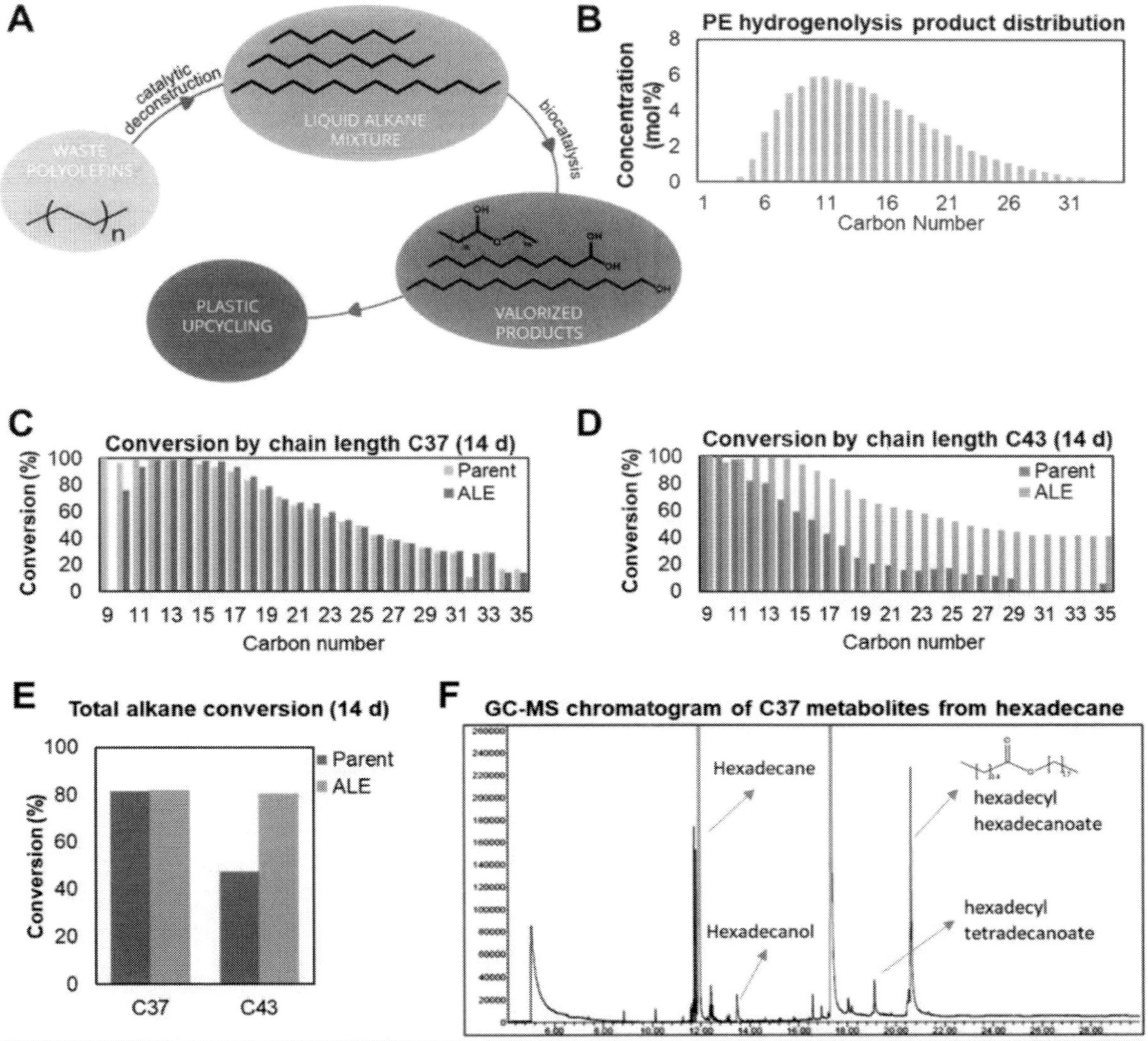

Figure 1. (A) Overarching goal of the project is to identify microbial consortia that can valorize polyolefin catalytic deconstruction products. **(B)** The liquid product distributions for the hydrogenolysis reaction of HDPE (MW= 76 kDa) over a carbon-supported catalyst. Carbon conversion of individual alkane chain lengths by environmental consortium **(C)** C37 or **(D)** C43 before and after adaptive laboratory evolution (ALE). **(E)** Quantification of PE deconstruction product conversion after microbial treatment for 14 days with C37 or C43 before and after ALE. **(F)** GC-MS chromatogram identifying metabolites produced by C37 after 14 d of growth on hexadecane.

473d Degradation of Polyethylene into Olefin Containing Products through Mechanochemistry

Laura Wilcox, Douglas P. Theberge, Yagnaseni Ghosh

Plastics have become widespread in their use with the majority of plastic being single use, thus, leading to approximately 300 million tons of plastic waste are generated annually. The recycling industry and infrastructure, however, is not equipped to handle this magnitude of plastic leading to only ~10-15% being recycled and the remaining waste being landfilled, incinerated, or polluted to the environment (i.e., oceans).[1, 2] In order to combat the plastic waste crisis, chemical recycling to convert waste into value-added products, such as hydrocarbon fuels and petrochemicals has been explored. Current methods include pyrolysis to convert plastic waste into hydrocarbon fuels by using high temperatures (500-1000 °C) in the absence of oxygen to thermally crack the polymer chains into lower molecular weight fractions.[3] The formation of these lower MW is implanted in refinery processes to convert to chemicals as well as may be used as transportation fuel. Although pyrolysis has resulted in production of olefins, the many drawbacks, including catalyst coking, high energy requirements, and a relatively large carbon footprint, limit the implementation of this technology. [4-8] To combat these challenges and drawbacks inherently present in the catalytic pyrolysis methods, here a different chemistry and process was explored known as mechanochemistry. Mechanochemistry refers to a chemical reaction that is induced by the direct absorption of mechanical energy.[9] Mechanochemical reactions are often solid-state carried out in mixer mills or planetary ball mills with the major advantages including operation without the use of solvents, reduced reaction times, and changes in product selectivity, and being less energy intensive.[10]

In this work, polyethylene (MW = 4000 da) was degraded using mechanochemical reactions catalyzed with aluminosilicate materials of varying properties such as Si/Al and topology (i.e., FAU and MFI zeolites). Mechanochemical reactions were conducted in an oxygenated environment at ambient temperature for up to 12 h using a planetary ball mill (Retsch PM 400) with stainless-steel jars containing grinding balls, polyethylene, and the aluminosilicate catalyst. Post reaction, a series of steps were taken to extract the formed olefin-containing residue from the catalyst and stainless-steel grinding balls including sonication, centrifugation, and rotary evaporating. Following the extraction of the residue, thermogravimetric analysis (TGA) was measured on the resulting product and revealed the product degraded faster and at a lower temperature than the reference polyethylene. ^{1}H, ^{13}C, and Heteronuclear Single Quantum Coherence (HSQC) Nuclear Magnetic Resonance (NMR) spectra were collected and revealed the presence of olefinic peaks in the range of ~5-5.3 ppm (^{1}H) and ~170-200 pm (^{13}C) with HSCQ allowing for a correlation between proton and carbon NMR spectra to determine the bond correlations. Overall, this low temperature, energy efficient mechanochemical process demonstrates that polymers can be successfully degraded to lower MW residues containing olefinic functional groups. This reveals the ability and effectiveness of mechanochemistry in degrading polymers into smaller constituent building blocks, including olefins and alkanes, thus paving way for greener and energy efficient solutions to breakdown polymers into monomers.

[1]Parker, L. *A Whopping 91 Percent of Plastic Isn't Recycled.* National Geographic Society, 2019.

[2]Programme, U.N.E., *Single-use plastics: A roadmap for sustainability.* 2018, United Nations Environment Programme.

[3]Jiang, J., et al., *From plastic waste to wealth using chemical recycling: A review.* Journal of Environmental Chemical Engineering, 2022. **10**(1).

[4]Li, Q., et al., *Progress in catalytic pyrolysis of municipal solid waste.* Energy Conversion and Management, 2020. **226**.

[5]Xue, Y., P. Johnston, and X. Bai, *Effect of catalyst contact mode and gas atmosphere during catalytic pyrolysis of waste plastics.* Energy Conversion and Management, 2017. **142**: p. 441-451.

[6]Wang, X., et al., *Soot formation during polyurethane (PU) plastic pyrolysis: The effects of temperature and volatile residence time.* Energy Conversion and Management, 2018. **164**: p. 353-362.

[7]Santos, B.P.S., et al., *Petrochemical feedstock from pyrolysis of waste polyethylene and polypropylene using different catalysts.* Fuel, 2018. **215**: p. 515-521.

[8]Huang, J., et al., *Chemical recycling of plastic waste for sustainable material management: A prospective review on catalysts and processes.* Renewable and Sustainable Energy Reviews, 2022. **154**.

[9]Horie, K., et al., *Definitions of terms relating to reactions of polymers and to functional polymeric materials (IUPAC Recommendations 2003).* Pure and Applied Chemistry, 2004. **76**(4): p. 889-906.

[10]Howard, J.L., Q. Cao, and D.L. Browne, *Mechanochemistry as an emerging tool for molecular synthesis: what can it offer?* Chemical Science, 2018. **9**: p. 3080-3094.

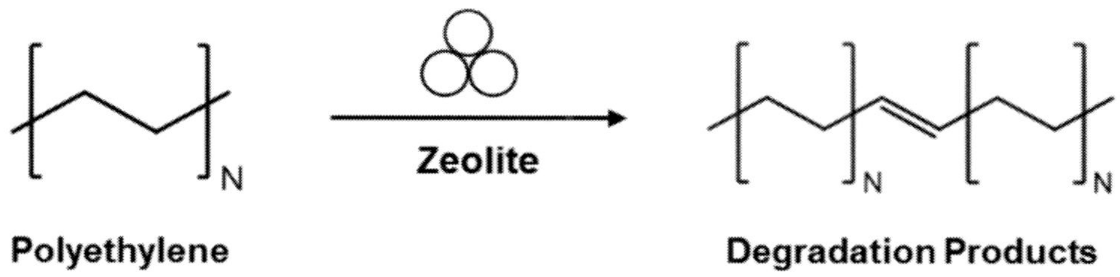

Figure 1. Mechanochemical degradation of polyethylene into olefin containing products using aluminosilicate catalysts.

473f A Study of Metal Supported ZSM-5 and Y Zeolite for Microwave-Induced Degradation of Low-Density Polyethylene

Thang Luong, Yuxin Wang, Sean Brown, Jianli Hu

Pyrolysis of plastic has been a topic of research because it was proposed as an alternative to landfilling and mechanical. This method is more energy efficient for conversion of plastics to high valued products such as olefins and aromatics. Recently, the use of microwave for degradation of plastic has been a novel technology to solve the waste plastic problem. The benefit of using microwave is its rapid and selective heating of catalyst. In this research, low density polyethylene was pyrolyzed using Fe promoted HZSM-5 catalyst in a microwave batch reactor. The reaction took place at 300°C for 30 minutes with N_2 flowing at 30 mL/min for an inert environment. Volatiles and condensate percentage were collected and compared between samples. The composition of these products was also investigated, with the gas composition for Fe supported zeolite Y and H-ZSM5 through time shown in the provided image. Overall liquid and gas composition will also be discussed. These profiles were compared to those reported in literature in terms of gas percentage with temperature ranging from 400-550°C. Plastic to catalyst ratio will also be varied to observed difference in the products. Based on product distribution results, interpretations can be made to maximize desirable products such as aromatics and olefins.

Unlike conventional heating, catalyst particles directly interact with the electric (or alternatively magnetic) field component of electromagnetic radiation, generating heat (so, the microwave heating is also known as dielectric heating). Thus, microwave heating occurs via two different mechanisms of dipolar polarization and Maxwell-Wagner polarization. It was observed that H-ZSM5 catalyst with and without some transition metal supported failed to reach the desired temperature due to low absorbance to microwave. A study on dielectric properties of the plastic and different catalysts will be conducted. Coupled with the selective heating, dielectric property can elucidate the heating mechanism of microwave on these materials.

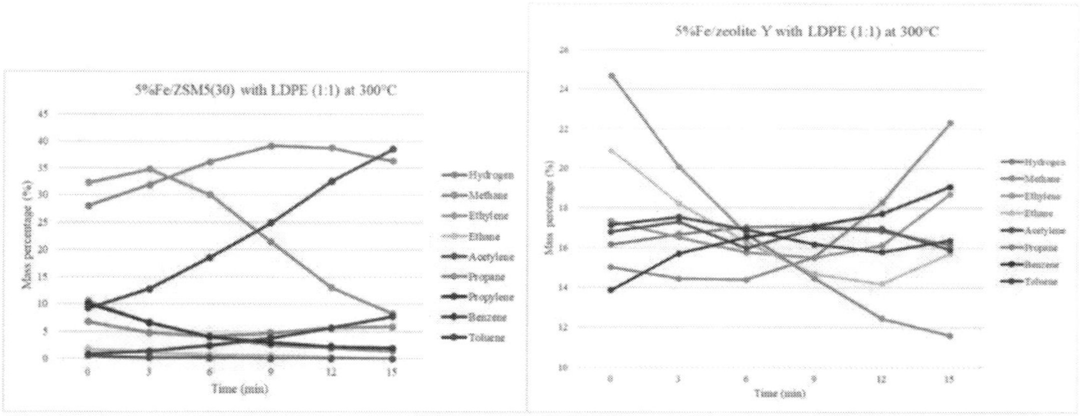

473g Impact of Acid Modified DBU-Based Ionic Liquids As Catalysts on Glycolysis of Poly (ethylene terephthalate)

Fahimeh Forouzeshfar, Maria Coleman, Joseph Lawrence

Polyethylene Terephthalate (PET) is one of the most popular packaging materials available worldwide for beverages. Even though mechanical recycling of PET is widely used in industry, due to the lack of availability of purer recycle streams, less percentage of PET waste is currently being recycled. Due to the increase in the generation of post-consumer PET, alternate recycling methods such as glycolysis is of interest. Glycolysis uses an excess of ethylene glycol as a solvent and reagent in the presence of a catalyst to break the ester bond and depolymerize the PET into Bis (2- hydroxyethyl) terephthalate (BHET). This BHET can then be converted back into PET. The aim of this study is to develop a series of 1,8-diazabicyclo [5.4.0] undec-7- ene (DBU) based salts as catalyst for the glycolysis of PET. The catalysts were developed by combining the super base DBU and a series of acids such as succinic acid, azelaic acid, and pimelic acid. The impact of reaction time, and catalyst/PET and EG/PET ratios on the conversion of PET and yield of BHET were monitored. The newly developed catalyst's effectiveness and selectivity in the production of oligomers and BHET were compared to DBU during glycolysis. Effect of the presence of the acid group on catalyst activity was studied. A possible reaction mechanism is proposed and confirmed using FTIR and NMR analysis. Finally, the catalysts' recoverability and reusability will also be discussed.

473h Microwave-Enhanced Catalytic Upcycling of Polymer

Yuxin Wang, Jianli Hu

Chemical upcycling of polymer represents an attractive approach of using polymeric waste as feedstock in promoting the circular economy and decarbonizing chemical industry. Here we report a one-step microwave catalytic approach for the deconstruction of plastic polymers into monomeric ethylene and BTX aromatics (benzene, toluene, and xylene) that can be used to re-construct polymers.

Currently, the majority of polymer ends up in landfills or the environment, harming the ecosystem and affecting the natural environment. Thermochemical pathways (e.g., pyrolysis, gasification) and high pressure hydrogenolysis enable depolymerization films to produce small molecules that could be used as fuel or integrated into chemical refineries. However, these thermochemical processes suffer from severe challenges which include inevitably higher energy input leading to higher greenhouse gas (GHG) emission (i.e., ~12 kg CO_2/1 kg H_2 production in gasification); poor product selectivity; low value of products, thus requiring significant upgrading for converting to valuable products.

Microwave catalytic process is a promising process for polymer upcycling with the advantages of lower reaction temperature, higher product selectivity, and simple one-step reaction process. Figure 1a presents the overall yields of LDPE under microwave and conventional thermal heating conditions at 300 °C. Under microwave catalytic upcycling conditions, gas yield reached 92% and the residue yield was only 4%. While, under conventional thermal heating, the gas yield was only 74% and the residue yield rose to 12%. Microwave is an efficient method for the polymer depolymerization. The difference of using microwave and conventional thermal heating methods is also reflected by the selectivity to product, particularly for the desirable BTX (Benzene, Toluene, and Xylene) and olefins productions. Comparing the products obtained on microwave and thermal, evident difference was obtained. As shown in Figure 1b, the $C_2^=$-$C_4^=$ and C_2-C_4 are dominant of over 95% products in gas on conventional thermal heating route, and the BTX selectivity is only 2%. This is mainly caused by the self-decomposition of plastic polymers by random chain scission mechanism, as the PE is heated up together with catalyst. Microwave catalytic upcycling gave primarily BTX with 45% selectivity, ZSM-5 can effectively convert the alkene intermediates to BTX at the same reaction temperature under microwaves. The catalyst exhibits two functions under microwave, catalysis function and the function of converting electromagnetic energy into thermal heat energy (Figure 1c). The heat is initially generated on the catalyst particles, which have good dielectric loss under microwave irradiation. In contrast, polymer remains cold because it is transparent to microwave (Figure 1d). The hypothesis is that the "hot spots" presented at catalyst-polymer interface facilitate the β–scission to form C_2 olefinic intermediates in cracking LDPE. Further catalytic reaction will depend on the catalytic properties of catalyst.

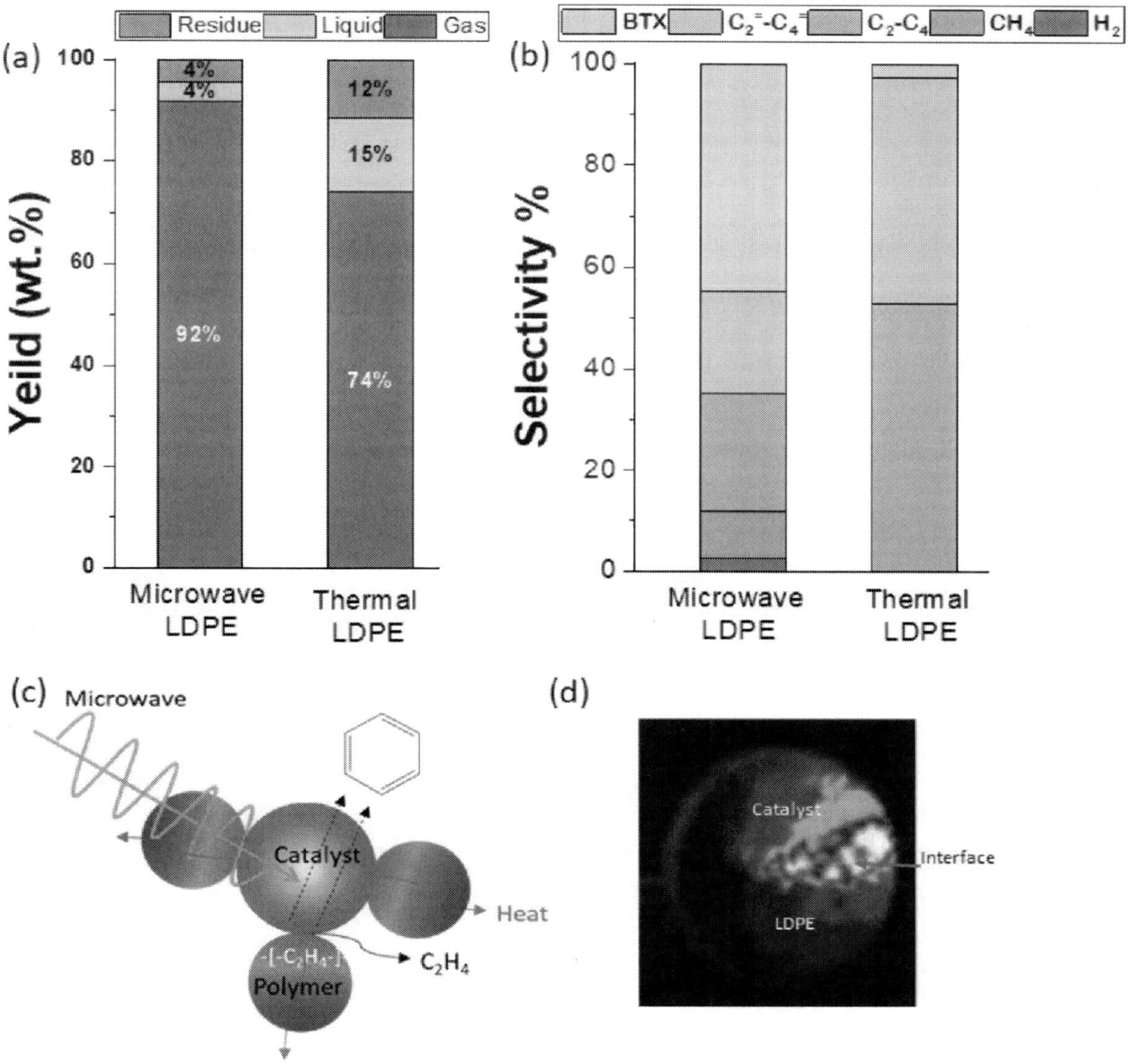

Residue
Liquid
Gas
(a)
Yeild (wt.%)
4%
4%
92%
12%
15%
74%
Microwave LDPE
Thermal LDPE
BTX
$C_2^=$-$C_4^=$
C_2-C_4
CH_4
H_2
(b)
Selectivity %
Microwave LDPE
Thermal LDPE
(c)
Microwave
Catalyst
Heat
-[-C_2H_4-]-
Polymer
C_2H_4
(d)
Catalyst
Interface
LDPE

473i Post-Consumer HDPE Waste Conversion to Lubricants: Techno-Economic Analysis and Life Cycle Assessment

Vincenzo Cappello, Pingping Sun, Guiyan Zang, Shishir V Kumar, Ryan Hackler, Hernán E. Delgado, Amgad Elgowainy, Massimiliano Delferro, Theodore Krause

Context

Global plastic production reached almost 370 million metric tons (MMT) in 2020. Most of this plastic ends up in landfills as waste at the end of its useful life with only a small fraction of the disposed plastics being recycled. Plastic upcycling to valuable chemicals could ensure a circular economy for plastics and reduce their environmental burden. In our study, we designed a process to convert high-density polyethylene (HDPE) to a high-quality liquid (HQL) product, which has tribological properties similar to premium lubricant oils such as Group III and PAO.[1]

Method

Based on experimental results for plastic hydrogenolysis using a novel catalyst technology that we developed, a large scale plant that would upgrade 250 MT/day of post-consumer HDPE to a high-quality lubricant-like product was designed. The process was modeled with Aspen Plus v12. Two different scenarios, a low-yield (LY) and a high-yield (HY) case, were studied. To compare the different scenarios, the minimum selling price (MSP) was determined (i.e., the production cost that would lead to a zero net present value). Life cycle assessment (LCA) was performed to evaluate the onsite and upstream emissions of the conversion process. Emissions were calculated using Argonne's GREET (Greenhouse Gases Regulated Emissions and Energy Use) LCA tool. The results were compared to the fossil fuel-based production process for conventional Group III and PAO lubricants.

Results and Conclusions

A techno-economic analysis (TEA) shows a breakeven production cost of $1.85 for the HY case and $6.03 for the LY case per gallon of HQL based on the operating conditions (Figure 1). For comparison, the average selling price for lubricant was $10.9/gal in 2020. The lowest production cost was obtained for the HY case with a 90% HQL yield. The HY case was characterized by a shorter reaction time and lower amount of catalyst usage compared to the LY case. At a MSP of $1.85/gal, the HDPE conversion has a greater margin than sorting and separately selling the plastic feedstock, converting it to liquid fuels via gasification and syngas processes or generating a synthetic crude via pyrolysis. Our conceptual design exhibited a lower cradle-to-gate CO_2 emissions compared to the conventional processes for producing Group III and PAO lubricants from fossil fuels. For the HY scenario, the onsite and upstream emissions were 0.33 $kgCO_2/kgHQL$. For comparison, emissions for conventional Group III lubricants is about 1 $kgCO_2/kg_{Lub}$, and for PAO is 1.65 $kgCO_2/kg_{Lub}$.[2,3] Thus, lubricants derived from plastic waste can reduce CO_2 emissions by 67% compare to Group III and 80% compared to PAO.

References

1 R. A. Hackleret al., *ChemSusChem*, DOI:10.1002/cssc.202100912.

2 P. Sun et al., *Environ. Sci. Technol.*, 2019, 53, 6556–6569.

3 Wang, Michael et al. *Greenhouse gases, Regulated Emissions, and Energy use in Technologies Model ® (2021 Excel)*, Argonne National Laboratory (ANL), Argonne, IL (United States), 2021.

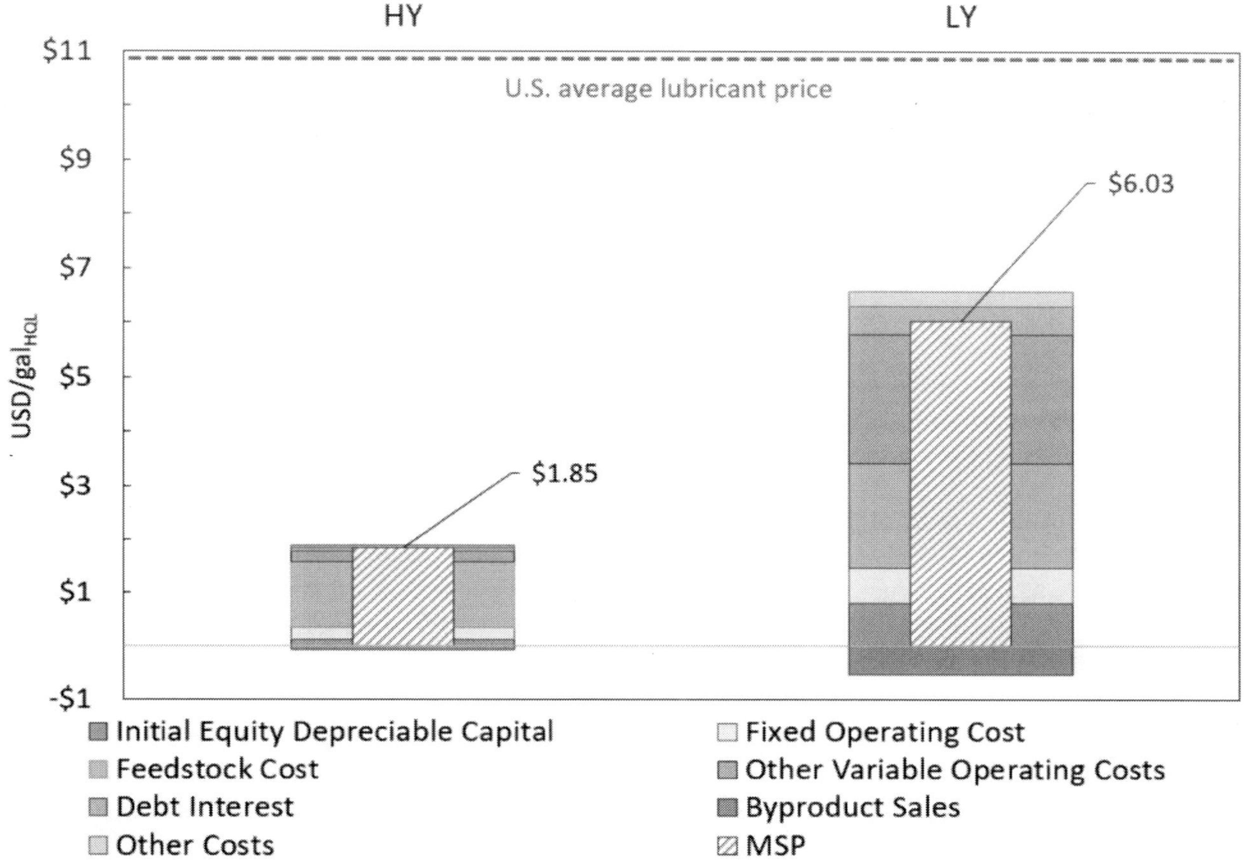

473j Environmental Benefits of the Solvent-Targeted Recovery and Precipitation (STRAP) Process for Multilayer Plastic Films Recycling

Aurora Del Carmen Munguia Lopez, Dilara Goreke, George Huber, Victor Zavala

Advances in the food packaging industry have brought diverse environmental and societal benefits; for instance, packaging is essential for enabling long shelf-lives, to mitigate food waste, and to increase food accessibility [1]. Food packaging often appears in the form of multilayer plastic films that leverage properties of different plastic materials to achieve different food protection goals (e.g., block temperature, light, and oxygen) [2,3]. Multilayer films help reduce packaging costs because, often, large amounts of single-layer materials are required to try to achieve the same properties [4]. Common plastics used in multilayer films include polyethylene (PE), ethylene vinyl alcohol (EVOH), and polyethylene terephthalate (PET). The films are commonly made of 2 to 17 layers of different polymers; however, some films can contain up to hundreds of distinct layers [5, 6].

Despite the many benefits of multilayer films, their production and use generate large amounts of non-biodegradable waste. In Europe, approximately 20% of food packaging is multilayer films; this proportion is expected to grow at an average of 9% per year due to current trends (e.g., convenience products, single-serve meals, and small packaging) [7]. Up to 40 % of the millions of tons of films produced end up as post-industrial waste due to inefficiencies and manufacturing errors. Traditional recycling methods (e.g., mechanical recycling) cannot recover each layer of the films due to their chemical incompatibility [8, 9]. Because of this, most film waste ends up in landfills, which is an environmental problem but, at the same time, a missed economic opportunity. Recently, a new technology called the STRAP process has been proposed as a promising alternative for recycling post-industrial film waste [10]. The process uses a series of solvent washes to separate the constituent polymers of different films. The first variation of this approach, STRAP-A, precipitates the polymers by adding antisolvents. A second process, STRAP-B, was proposed to reduce the use of antisolvents by leveraging the temperature-driven precipitation. Both approaches were able to deconstruct a post-industrial multilayer film composed mainly of PE, EVOH, and PET. A third process, STRAP-C, also leverages temperature-driven precipitation and was designed to treat a multilayer film of a different composition [11].

The environmental impacts and benefits of the STRAP processes over virgin resins production are highly relevant for implementing this technology commercially. Thus, this work evaluates the impacts on global warming potential, energy demand, water usage, and process toxicity using a life-cycle assessment methodology. The process is analyzed considering a product perspective [12]. From this perspective, the STRAP approach is seen as an alternative process to produce virgin-grade polymers (raw material of multilayer films). The functional unit considered is the production of 1 kg of multilayer film; therefore, we compare the environmental impacts of producing 1 kg of film from virgin resins with producing 1 kg of film from the recovered resins through the STRAP processes. We find that the STRAP-B and STRAP-C processes can help achieve the goals of a circular economy by reducing resource consumption and environmental impacts.

References:

[1] M. Niaounakis. Recycling of flexible plastic packaging. William Andrew, 2019.

[2] A. S. Bauer, M. Tacker, I. Uysal-Unalan, R. Cruz, T. Varzakas, and V. Krauter. Recyclability and Redesign Challenges in Multilayer Flexible Food Packaging—A Review. Foods, 10(11):2702, 2021.

[3] K. Kaiser, M. Schmid, and M. Schlummer. Recycling of polymer-based multilayer packaging: A review. Recycling, 3(1):1, 2017.

[4] S. Ebnesajjad. Plastic films in food packaging: materials, technology and applications. William Andrew, 2012.

[5] T. Anukiruthika, P. Sethupathy, A. Wilson, K. Kashampur, J. A. Moses, and C. Anandharamakrishnan. Multilayer packaging: Advances in preparation techniques and emerging food applications. Comprehensive Reviews in Food Science and Food Safety, 19(3):1156-86, 2020.

[6] J. R. Wagner Jr. Multilayer flexible packaging technology and applications for the food, personal care, and over the counter pharmaceutical industries. William Andrew, 2009.

[7] A. Grant, L. Lugal, and M. Cordle. Flexible Films Market in Europe: State of Play. Eunomia: London, UK, 2020.

[8] S. Billiet and S. R. Trenor. 100th anniversary of macromolecular science viewpoint: needs for plastics packaging circularity. ACS Macro Letters, 9(9):1376-90, 2020.

[9] O. Horodytska, F. J. Valdés, and A. Fullana. Plastic flexible films waste management–A state of art review. Waste management, 77:413-25, 2018.

[10] T. W. Walker, N. Frelka, Z. Shen, A. K. Chew, J. Banick, S. Grey, M. S. Kim, J. A. Dumesic, R. C. Van Lehn, and G. W. Huber. Recycling of multilayer plastic packaging materials by solvent-targeted recovery and precipitation. Science advances, 6(47):eaba7599, 2020.

[11] K. L. Sánchez-Rivera, P. Zhou, M. S. Kim, L. D. González Chávez, S. Grey, K. Nelson, S. C. Wang, I. Hermans, V. M. Zavala, R. C. Van Lehn, and G. W. Huber. Reducing Antisolvent Use in the STRAP Process by Enabling a Temperature-Controlled Polymer Dissolution and Precipitation for the Recycling of Multilayer Plastic Films. ChemSusChem, 14(19):4317-29, 2021.

[12] H. Jeswani, C. Krüger, M. Russ, M. Horlacher, F. Antony, S. Hann, and A. Azapagic. Life cycle environmental impacts of chemical recycling via pyrolysis of mixed plastic waste in comparison with mechanical recycling and energy recovery. Science of the Total Environment, 769:144483, 2021.

473k Production of Macronutrients from Hybrid Thermal and Biochemical Conversion of Waste Plastic

Jessica Brown, Efrain Rodriguez-Ocasio, Tannon Daugaard, Ryan Smith, Laura R. Jarboe, Robert Brown

Although plastics are thought not to be biodegradable, plastics are actually broken down very slowly by microorganisms in the environment. To improve upon the long time-scales of plastic degradation, the natural abiotic and biotic processes that synergistically decompose plastics can be accelerated. Using thermal oxo-degradation, abiotic deconstruction becomes rapid as polymer backbones are cracked while adding oxygen functionality to fragmented products. A mixture of hydrocarbons, alcohols, aldehydes, and carboxylic acids are produced from polyolefins subjected to moderate temperatures in an oxidative environment. The added oxygen functionality dramatically expedites yeast metabolism of the degraded plastic without supplemental carbon, as growth of yeast is observed in 48 hours. As yeasts use the degraded plastic as a carbon source on which to grow, a pathway to breakdown waste plastic for biological upgrading to macronutrients in the form of microbial biomass is possible.

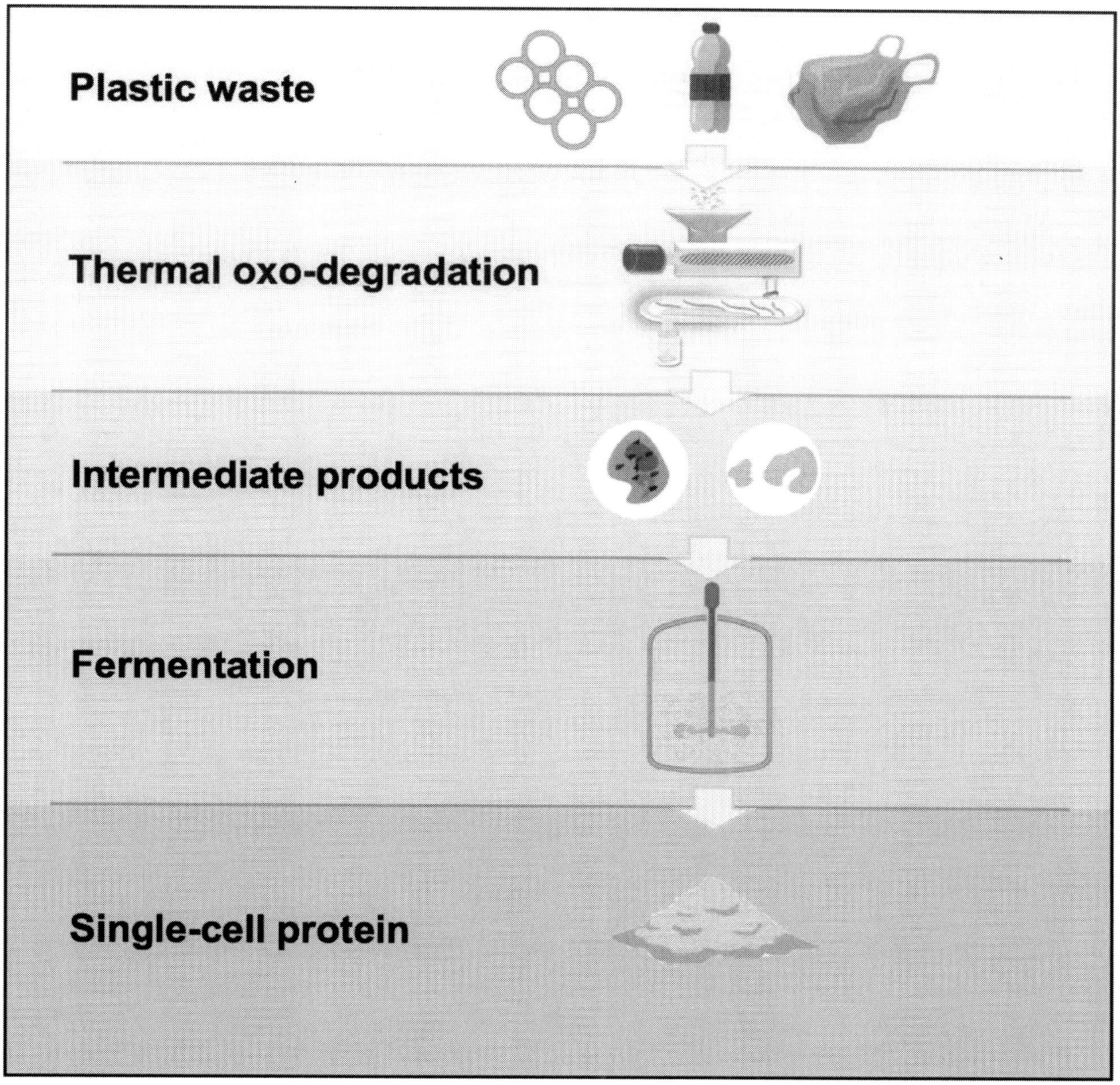
Plastic waste
Thermal oxo-degradation
Intermediate products
Fermentation
Single-cell protein

473m Implement a Circular Ecomony for Plastic Waste in Developing Countries

Jeffrey Seay, Dimitrios Karadimas, Shelby Browning, Kaitlyn McGlennon

Waste plastic is a significant global challenge, especially in infrastructure limited countries. However, using appropriate technology to convert plastic waste into fuel oil, a locally managed, decentralized circular economy can be established. Providing an economic value to waste plastic can benefit the environment as well as the local community. Novel research conducted in collaboration with Makerere University in Uganda and the non-profit organizations Empowered Solutions for Environmental Sustainability and Upcycle Africa focused on implementing a circular economy for plastic waste in Uganda will be presented.

AUTHOR INDEX